Reviews of A Brief Guide to Art

I0014463

"An excellent short primer to quickly get the uninitiated grounded in the field of artificial intelligence."

Professor Eric Topol, MD, Executive Vice-President, Scripps Research, USA.

"There are very few people that have both a technical understanding of recent achievements in artificial intelligence, and a deep appreciation of the nature of human intelligence. James Stone is one of those people. Fortunately for us, he has distilled much of his understanding into this book, which doesn't sacrifice accuracy for brevity, and which summarises recent achievements, as well as the challenges that lie ahead. I'm often asked to recommend a short introduction to artificial intelligence; this book will do the job perfectly."

Neil Lawrence, DeepMind Professor of Machine Learning, University of Cambridge, and Senior AI Fellow, Turing Institute.

"This is the book you need if you want to understand the basics of AI, including its triumphs and limitations. Stone takes the reader on an authoritative whirlwind tour, and delves into just enough technical detail to give an overview of key concepts, such as neural networks, deepfakes, and overfitting. An essential and timely summary of modern AI."

James Marshall, Professor of Computational Biology, University of Sheffield, and Co-Founder, Opteran Technologies.

James V Stone is an Honorary Associate Professor at the University of Sheffield, UK.

A Brief Guide to
Artificial Intelligence

James V Stone

Title: A Brief Guide to Artificial Intelligence
Author: James V Stone

©2020 Sebtel Press

All rights reserved. No part of this book may be reproduced or transmitted in any form without written permission from the author. The author asserts his right to be identified as the author of this work in accordance with the Copyright, Designs and Patents Act 1988.

First Edition, 2020.
Typeset in LATEX 2_ε.
First printing.

ISBN 9781916279117

Cover background modified from an original image by Okan Caliskan, reproduced with permission.

For sentient beings, both biological and artificial

Contents

Preface

There are plenty of long books about how artificial intelligence (AI) works, and there are many extensive books about why AI matters. In contrast, this is a short book that summarises both how AI works and why it matters.

Given that this is a short book, it is only fair to state not only what is included but what is omitted, because novice readers can judge the former but not the latter. Readers who want to learn the essentials of AI do not want, or need, to know the minutiae of every AI software package. Nor do they need to know the mathematical theory behind any particular learning algorithm. Accordingly, these details are not included. What is included is a comprehensive overview of current AI methods, how they work, their applications, and their limitations.

The Author. In 1984, Geoffrey Hinton (one of the founders of modern AI) gave a lecture on an obscure form of artificial neural networks called *Boltzmann machines*. I was so inspired by this lecture that I implemented a Boltzmann machine for my MSc in Knowledge Based Systems, and I later enjoyed an extended research visit to Hinton's laboratory in Toronto. With a PhD in Computer Vision, I have published ten books and more than one hundred scientific papers on AI, vision, information theory, and computational neuroscience.

Note that a small proportion of the text here is based on the non-mathematical sections of my book *Artificial Intelligent Engines*.

Acknowledgements. For reading one or more chapters, I am grateful to Nikki Hunkin, Lancelot Da Costa, Stephen Eglen, John Frisby, Tom Peretz, Maya Williams-Hamm, and Stuart Wilson. Thanks to Amealia Wharmby for the picture of her dog in Figure 1.2. Finally, thanks to Alice Yew for meticulous copy-editing and proofreading.

Corrections. Please email corrections to
j.v.stone@sheffield.ac.uk
A list of corrections can be found at
https://jim-stone.staff.shef.ac.uk/AIGuide

Once the machine thinking method has started, it would not take long to outstrip our feeble powers.
Alan Turing, 1951.

Chapter 1

What Is Artificial Intelligence?

1.1 Introduction

Artificial intelligence (AI) is the ability of machines to behave intelligently, either in the form of computers or as robots. This immediately raises the question: exactly what do we mean by intelligent behaviour? Even though most of us think we can spot intelligent behaviour when we see it, we would be hard pressed to come up with a rigorous definition. This question has occupied psychologists and ethologists for over a century, and the resultant debates often generate more heat than light. To avoid such intellectual quagmires, we will make use of a simple informal definition here:

> *Intelligence is the ability to solve problems with a small number of attempts at a solution.*

To see how this definition applies to modern AI systems, consider two examples, one of partial success and one of failure. The example of partial success is an AI system that learned to play chess better than any human. The success is partial because learning occurred over more games than a human could play in a lifetime. In contrast, no AI system or robot can yet perform the relatively mundane task (for a human) of building a wardrobe that has arrived in a flatpack box (i.e. unpack it, read the instructions, and build the wardrobe).

These examples suggest that current AI systems are extremely good at a narrow range of tasks (e.g. image recognition and game playing) but appallingly bad at most everyday tasks (e.g. wardrobe construction). It could be argued that chess playing and wardrobe construction require two different types of intelligence, but this kind of argument inevitably leads to a fractionation of intelligence that has no obvious end point (e.g. would we require one type of intelligence for building wardrobes and another for building desks?).

1.2 AI, AGI and Other Euphemisms

The debate over whether modern AI systems are intelligent has resulted in two typical responses. The first is to acknowledge that the term AI does not accurately describe the narrow capabilities of modern systems, and to define the broader spectrum of skills required of true intelligence as *artificial general intelligence*, or AGI. However, this does not really help because it just postpones the problem of how to define AGI. Inevitably, any such definition will prove to be inadequate, so it will be necessary to generate new euphemisms, such as *real artificial general intelligence* (RAGI); thankfully, this term does not exist, yet.

The second response is to declare that anything a computer (or robot) can do is not intelligent behaviour. This provides a simple definition, but one that is both scathingly cynical and extraordinarily unhelpful: *AI is what computers cannot do (yet).* However, this tautological definition should not be dismissed out of hand, because

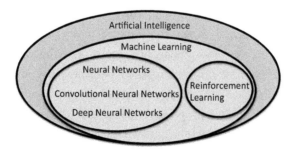

Figure 1.1: The main categories of artificial intelligence.

(although they would not admit it) it seems to be widely employed by the most ardent critics of AI systems.

So, are modern AI systems intelligent? According to the definition of intelligence given on page 1, the answer must be *no*. However, this does not mean that AI systems in the future will not be intelligent (see Chapter 4). Even though current AI systems are not intelligent, the term *artificial intelligence* has become synonymous with almost any sophisticated computer program (and occasionally also with some extraordinarily pedestrian programs). In order to avoid clumsy phrases like 'modern so-called AI systems', the term AI will be used throughout this book to refer to current systems based on any form of machine learning (see Figure 1.1, and see the Glossary for definitions).

Naming Is Not Knowledge: AI, ML and DL. While we are on the topic of naming things, we should consider the plethora of other terms associated with AI. Quite justifiably, novices frequently ask for precise definitions of the terms AI, *machine learning* (ML), *artificial neural network* (ANN), *deep learning* (DL), and *deep neural network* (DNN), each of which is described in the Glossary. In fact, there are no hard borders between these terms, which exist mainly as a result of historical accident. Having said that, the widespread use of these terms means that we should have some idea of how they are related. To this end, a rough guide is provided in Figure 1.1.

1.3 So, What Is AI?

So, what is AI? This will be answered in more detail in subsequent chapters, but a brief answer is as follows. Currently, AI is dominated by various forms of *artificial neural networks*, like the one in Figure 1.2. The algorithmic engines that drive most modern AI systems involve some form of the *backpropagation*, or *backprop, algorithm*, which is often combined with the *reinforcement learning algorithm*. In essence, the backprop algorithm is used for learning input/output pairs (as in Figure 1.2), whereas reinforcement learning is used for learning sequences of actions (e.g. playing chess; see Section 2.4).

The term algorithm is often used as if the word itself has magical properties, but an algorithm is just a formal recipe involving a definite sequence of steps that are guaranteed to yield a particular result. For example, in the game of tic-tac-toe, or noughts-and-crosses as it is known in the UK, it is not hard to write down the optimal next move for each state of a game. This description is, to all intents and purposes, an algorithm, which can easily be converted into a computer program, so that a computer can then play tic-tac-toe. Admittedly, such a program would not enable the computer to *learn* to play, but a *learning algorithm* for the game could be written down and then implemented as a computer program. By running the resultant program, the computer could then learn to play. This program would be simpler than those used in AI systems, but no different in principle.

Without delving into the long history of attempts to make computers behave as if they were intelligent, the modern era of AI

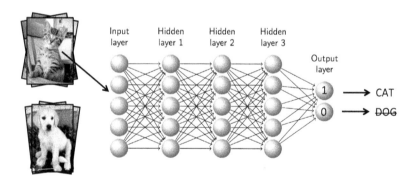

Figure 1.2: Classifying images into two categories, *cat* and *dog*, using a deep neural network (DNN) with five layers of artificial neurons or *units*. The *layers* between the input and output layers are called *hidden layers*. Each unit is connected to every unit in the next layer by a *weight*. Weight values are learned gradually (using the backprop algorithm) during repeated exposure to many input/output pairs (e.g. a picture of a cat and the word *cat*). Here, the input is a picture of a cat, and (in response) the output unit representing the word *cat* is on (taking value 1), whereas the output unit representing the word *dog* is off (taking value 0).

can be traced back to around 1980. At that time, the idea of AI was dominated by *symbolic AI*, or *good old-fashioned AI* (GOFAI) as it is sometimes known. This was based on the assumption that computer programs consisting of high-level symbols and formal logic (if–then statements) would suffice to produce intelligent behaviour, usually without any form of learning from experience. However, a small, and until recently widely disregarded, group of researchers believed that AI could be achieved by copying the physical microstructure of the brain. And, because the brain consists of about 86 billion neurons, these researchers programmed computers to model small networks of neurons, called *artificial neural networks* (Figures 1.2 and 1.3).

Of the many problems implicit in this approach, the main one was that no-one knew of an algorithm for making artificial neural networks learn, even for small networks. After several false starts, and so-called *neural network winters*, a few key algorithms started to emerge (see Further Reading). With the help of Moore's law (which predicts a doubling of computing power approximately every two years), by about 2012 these algorithms began to make serious inroads into tasks such as game playing and object recognition (Figure 1.2) that require a degree of skill. Artificial neural networks are now so ubiquitous that they are usually referred to simply as *neural networks*.

1.4 Almost-Human Memory

Much of the power of neural networks depends on three key properties, which are also vital features of biological neural networks.

First, unlike conventional computer memory, neural network memory is *content addressable*. This means that recall is triggered by an image or a sound, analogous to the way a piece of music can bring to mind a particular memory. In contrast, computer memory can be accessed only if the location of the required information is known.

Second, a common by-product of content addressable memory is that given a learned association between an input image and an output, recall can be triggered by an input image that is merely similar to the original input of a particular input/output association. For example,

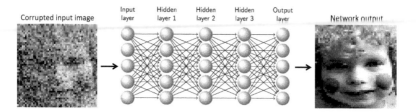

Figure 1.3: A neural network was trained to reproduce at its output layer an image presented to its input layer. Subsequently, a corrupted image (left) is presented to the input layer, and the 'response' (right) essentially recalls the original image. In practice, the number of units in the input and output layers would be the same as the number of pixels in the image.

a network can be trained to output the name of an object presented to its input as in Figure 1.2, even if the input image is corrupted by noise. This ability of neural networks to work with corrupted inputs can be seen in Figure 1.3. Here, a network has been trained to reproduce at the output layer the image presented to its input layer. When a corrupted version of an image is presented to the input layer, the output layer still produces the original image. This ability to *generalise* beyond learned associations is a critical property of neural networks.

Third, destroying a single weight or unit in a neural network does not destroy a particular learned input/output association; instead, it degrades all associations to some extent. This *graceful degradation* also occurs with human memory.

1.5 Interpretability

A common criticism of neural networks is that they are essentially *black box* systems. In other words, because a neural network learns from data, and because the knowledge gained by the neural network is implicit in the millions of connections between thousands of units, it is extremely difficult to know exactly how the network makes decisions regarding any particular input. There are two broad responses to this problem of *interpretability*. The first argues that it *is* possible[20] to interpret the decisions made by a neural network.

The second response is that such a criticism is simply unfair. When humans become expert in a particular domain, they can rarely explain why they took a particular decision[45]. Or rather, they can explain, but their explanations may have little to do with the actual process by which they made that decision[6]. This is because knowledge gained over many years becomes almost like instinct, so that the expert has little insight into why they took certain actions.

In fact, even though we do not recognise it as such, we all have expertise in everyday domains such as face recognition, language, and social interactions (well, most of us do). Despite the wealth of knowledge implicit in such expertise, we cannot easily articulate how we recognise a friend, or know when (or if) to laugh at a joke, any more than we can explain how to ride a bike or how to chew gum without biting our own tongue. Thus doctors, pilots, car drivers, and anyone with expertise in any domain are, to all intents and purposes, as opaque as AI systems.

The successes of AI systems have been tempered by the growing realisation that describing how the components of a 'black box' system work is not the same as understanding how the system as a whole works. This realisation was foreshadowed by Gottfried Leibniz (1646–1716), the philosopher and co-inventor (with Isaac Newton) of calculus:

> *Let us pretend that there was a machine, which was constructed in such a way as to give rise to thinking, sensing, and having perceptions. You could imagine it expanded in size ..., so that you could go inside it, like going into a mill. On this assumption, your tour inside it would show you the working parts pushing each other, but never anything which would explain a perception.*

Thus, opacity with respect to the physical mechanisms that instantiate intelligence may be an inevitable property of all intelligent systems, both artificial and biological.

1.6 Why Now?

Given the limitations of AI systems outlined above, why is there suddenly so much interest in AI? Because around the year 2012, neural networks started to outperform conventional computer vision systems in annual image recognition competitions (see Figure 2.1).

Not only did neural networks beat the best computer vision systems, they soon began to outperform humans on a range of tasks, most notably at chess. More importantly, unlike conventional AI systems, neural networks *learned*, often with minimal supervision, to outperform humans. And if neural networks can learn to recognise objects and play chess, what other tasks might they excel at? This question became a spur to action, giving rise to the recent achievements of AI.

These achievements are covered in detail in later chapters, but it is the fact that neural networks beat the best human players at sophisticated games such as chess and Go which finally caught the world's attention. Once companies including Google, Amazon and Facebook saw the potential benefits of AI, research laboratories quickly expanded and new laboratories were established. These well-funded laboratories attracted a tidal wave of talented individuals. As a result, the number of research papers produced on a daily basis could easily supply a ticker-tape parade.

Most of the flurry of papers that followed AI's successes described small, incremental changes that improved performance on a specific task by a few percent. But a substantial minority of those papers described novel and useful neural networks. These were the AI systems that began to make headlines, not only in renowned scientific journals such as *Nature* but also in newspapers and magazines. These AI systems were ground-breaking not because they were almost as good as human experts at playing games and diagnosing medical conditions, but because they were *better than any human*.

Chapter 2

What AI Can, and Cannot, Do

2.1 Introduction

AI systems based on neural networks are extraordinarily good at two types of tasks:

a) pattern recognition (e.g. face recognition, medical diagnosis);

b) performing sequences of actions (e.g. playing ping-pong).

This may not sound too impressive, but pattern recognition can be used to spot tumours in mammograms and to differentiate between benign skin blemishes and skin cancer. Moreover, as speech is basically a temporal pattern, it should come as no surprise that the speech recognition algorithms in smartphones and digital assistants (e.g.

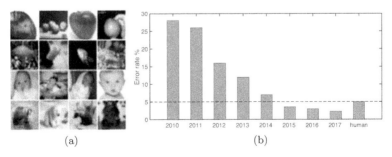

Figure 2.1: Classifying images. (a) Example images. (b) The error rate in an annual image classification competition has fallen dramatically since 2010. Reproduced with permission from Wu and Gu (2015).

Alexa) also rely on neural networks. Additionally, the ability to perform a sequence of actions to reach a particular goal in a game (e.g. chess) has been harnessed to improve the efficiency of power plants.

A promising general application of neural networks involves detailed simulation of complex physical systems. For example, conventional models of weather systems, galaxies and protein folding[2] require enormous amounts of computing time and power. Recently, neural networks have been shown to improve results while using a mere fraction of the resources needed for conventional models[16;33].

2.2 Visual Recognition

Some idea of the rapid progress of AI can be gleaned from the annual object recognition competitions held to assess the current state of computer vision systems. These competitions involve classifying about 1.5 million images into 1,000 object classes. As shown in Figure 2.1, between 2010 and 2014 performance improved dramatically, with the mis-classification rate falling from about 28% to 7%, which is close to that (5%) of humans. Crucially, from about 2015, AI systems based on neural networks not only beat conventional computer vision systems but also surpassed human levels of performance. More recently, this ability of AI systems to process images has been harnessed for face recognition and medical diagnosis (see Section 2.6).

Identifying Handwritten Letters. An important testing ground for neural networks is recognition of handwritten letters and numbers. The standard database for testing neural networks on this task is the MNIST data set, which contains 60,000 training images and 10,000

Figure 2.2: Images of handwritten numbers from the MNIST database.

test images, a few of which are shown in Figure 2.2. These types of data represented a serious challenge for neural networks in the 1980s, but the advent of *convolutional neural networks* (see Chapter 3) initiated a programme of work that has yielded rapid improvements in performance on the MNIST database. Today, it is not unusual for a network to have error rates of less than 0.5% (i.e. 99.5% of the test images are correctly classified).

Face Recognition.   One reason that face recognition is hard is the same reason that recognising any object is hard: objects can appear at any angle, any size or (almost equivalently) any distance (Figure 2.3). But faces, unlike most objects, also have a multitude of expressions, which presents a challenge even for some humans. Worse, the same face can have a bewildering variety of hair styles, make-up, glasses, beards, hats, and so on. Despite these challenges, in 2015 a deep convolutional neural network called FaceNet[38] was trained on 200 million face images and achieved an accuracy of 99.63%, which was a record at the time.

(a)   (b)

Figure 2.3: The face in (a) and in (b) is recognised as being the face of the same person by FaceNet[38] despite different viewing angles and different lighting conditions. Reproduced with permission from Gary et al. (2007).

2.3 Deepfakes

Once a neural network has been trained on faces, it is possible to get that network to generate entirely synthetic faces. For example, a type of neural network called a *generative adversarial network* (GAN) can generate faces that are combinations of the real faces it has been trained on. Consequently, the faces generated look realistic, but they are not the faces of anyone who actually exists[19] (Figure 2.4).

If a neural network can generate synthetic images of non-existent people, then it is a relatively small step to get such a network to generate a movie of a particular person speaking. Furthermore, if this network is given access to a large number of short movie clips of that person, it can generate clips in which the person speaks sentences that they never actually uttered[53]. A well-known example of such a *deepfake* video consists of Barack Obama apparently saying things that are, to say the least, uncharacteristic[27].

2.4 The Player of Games

The ability to learn to play games such as chess is not, in itself, of practical importance. But winning games matters because it demonstrates an ability to learn both short-term tactics and long-term strategies. As early as 1995, a *reinforcement learning algorithm* was used with a neural network to learn backgammon, and the resultant system was called *TD-Gammon*. Learning was achieved by duplicating the neural network program and then having the two resultant clones

Figure 2.4: Synthetic celebrities. These pictures were produced by a generative adversarial network that was trained on 30,000 images of celebrities. Reproduced with permission from Karras et al. (2018).

play against each other until one of them emerged as the winner. The experience gained during each game was used by the reinforcement learning algorithm to make changes to the neural network weights of each clone to improve its performance. These weight changes were then copied back to the original neural network, which therefore benefited from the experiences of both the winner and the loser, in readiness for the next game. After repeating this process for 1.5 million games, the TD-Gammon neural network played so well that it beat the best backgammon player in the world[52].

An intriguing aspect of the TD-Gammon neural network is that it developed a style of play that was unlike the repertoire of strategies used by human players. Indeed, the strategy developed by the TD-Gammon neural network was so novel, and so successful, that it was subsequently adopted by backgammon grandmasters.

Success at classic arcade-style Atari games was achieved in 2013 by a combination of deep learning and reinforcement learning. This resulted in superhuman levels of performance in games such as pong, which is a computer version of table tennis[29][30], shown in Figure 2.5.

A particularly impressive breakthrough was made in 2016, when an AI system learned to play the game of Go. This game has about 10^{170} legal board positions, more than the number of atoms in the known universe. The AI system was called *AlphaGo*[43], and it beat some of world's best human players in 60 out of 60 games. After being beaten three times, the world champion Ke Jie said, "Last year, I think the

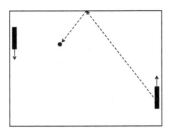

Figure 2.5: The game of pong is a computer version of table tennis. Each player can move one bat vertically to hit the ball. The arrow on each bat shows the direction of the current movement.

way AlphaGo played was pretty close to human beings, but today I think he plays like the God of Go."

By October of 2017, another AI system, *AlphaGo Zero*, learned to play Go at a level that was not only better than any human on Earth but also better than AlphaGo. Whereas AlphaGo initially learned by observing 160,000 human games, AlphaGo Zero learned through only trial and error before beating AlphaGo 100 games to none[44]. Like TD-Gammon, both AlphaGo and AlphaGo Zero relied on a combination of reinforcement learning and deep learning. Admittedly, the computers used to implement AlphaGo Zero were so powerful that the AI system could learn from playing many millions of games in a short time; nevertheless, AlphaGo Zero did learn to play games at superhuman levels of performance, and without any intervention from human players.

Finally, a stripped-down version of AlphaGo Zero, called *AlphaZero*, was found to learn faster than AlphaGo Zero. Remarkably, without altering the pre-learning parameter values, the algorithm that enabled AlphaZero to learn Go was then used to learn chess[39]. The best traditional computer chess program, *Stockfish*, already played at superhuman levels of performance. In a tournament of 100 games between Stockfish and AlphaZero, 72 were a draw, and AlphaZero won the remaining 28. As was the case for TD-Gammon and AlphaGo Zero, some of the moves made by AlphaZero seemed strange, until it was realised that those moves were instrumental in winning the game.

Just as TD-Gammon altered the strategies used by humans to play backgammon, so AlphaZero is changing the strategies that humans use

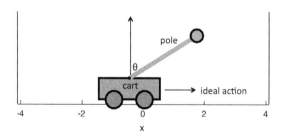

Figure 2.6: Learning to balance a pole on a cart.

to play Go; and it is entirely possible that AlphaZero will do the same for chess. So, in a sense, humans are starting to learn new strategies from machines that learn.

The importance of AlphaZero beating AlphaGo Zero, which beat AlphaGo, which beat the human world champion, cannot be overstated. We can try to explain away the achievements of AlphaGo Zero by pointing out that it played many more games than a human could possibly play in a lifetime, but the fact remains that a computer program learned to play from scratch so well that it can beat every one of the 7.7 billion people on Earth.

2.5 Skilled Control

The ability to learn short-term tactics and long-term strategies for games has been harnessed to learn real-world tasks that require substantial amounts of skill. An early benchmark for reinforcement learning is a task similar to balancing a pencil on the tip of your finger; the actual task is balancing a pole mounted on a cart by moving the cart from side to side. This pole-balancing task has been solved using reinforcement learning in simulated physical environments, as well as with real poles on real carts (see Figure 2.6 and Section 3.8). Even more impressive is using reinforcement learning

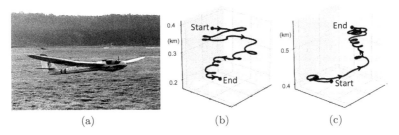

(a) (b) (c)

Figure 2.7: Learning to soar using reinforcement learning.
(a) Glider used for learning.
(b) Before learning, flight is disorganised; maximum height is 0.4 km.
(c) After learning, flight is coherent; maximum height is 0.6 km.
Figures based on Guilliard et al. (2018), reproduced with permission.

to fly. First, in a simulated environment, a glider learned to gain height by circling around thermals. With this encouraging initial development, reinforcement learning was then used in a physical glider with considerable success, as shown in Figure 2.7.

Driverless Cars. Given the ability of neural networks to recognise objects, it is only natural that neural networks have been harnessed for use in driverless cars. However, as any driver knows, there is more to driving than recognising objects such as people and traffic signs. Despite enormous amounts of research over the past few years[4], no self-driving car has yet been proved safe enough for general use.

2.6 AI in Medicine

Given that AI systems can learn to classify thousands of different objects, it is natural to ask whether they can be used to identify different types of cancer. This is particularly relevant to cancers for which diagnosis relies heavily on some form of image data, for example from CAT scans or X-rays.

In a landmark study[10] on skin cancer in 2017, the classification performance of a neural network trained on 130,000 images of skin blemishes was as good as that of dermatologists. Similarly, a study from 2018 found that neural networks can identify anomalies in mammograms just as well as radiologists[21]. When trained on genetic data, a deep neural network achieved an accuracy of 91% in identifying the type of cancer, which is about twice as accurate as pathologists presented with a metastatic tumour[17].

The ability of neural networks to identify pneumonia from chest X-rays alone was found to be better than the combined efforts of four radiologists[56]. Neural networks have been particularly useful in opthalmology, probably because diagnosis of retinal diseases depends almost exclusively on images of the retina. In 2020, a neural network[58] was trained on a combination of retinal (fundus) images and genetic data to predict progression of age-related macular degeneration (AMD), yielding a score of 0.85 on a performance measure called AUC

(an AUC of 1 corresponds to perfect performance; see Glossary). In a related application, the motion of a beating heart was used as input to a neural network, which gave substantially improved predicted survival rates for patients with right ventricular dysfunction[5].

Of course, diagnosing any disease involves more than simply looking at images, so results like these do not mean that clinicians will be replaced by neural networks. However, they may mean that, to quote Professor Langlotz[24] (a radiologist at Stanford University), "AI won't replace radiologists, but radiologists who use AI will replace radiologists who don't."

More generally, studies like these have led to growing acceptance of the role of AI in medical applications. As of 2019, 14 AI systems had been approved by the U.S. Food and Drug Administration[54] (FDA). By February 2020, the FDA had approved 50 algorithms. Furthermore, this number is set to increase rapidly, facilitated by the FDA's fast-track approval plan for medical AI systems, published in 2018. However, FDA approval is not always based on the gold standard for medical research studies, which is the *randomised controlled trial* (RCT). Indeed, fewer than 10 of the 50 FDA-approved applications have been tested in RCTs, and only 20 have even been tested in a clinical environment.

In vitro fertilisation (IVF) famously has a low success rate, so anything that improves matters is to be welcomed. A key indicator of success is development of the embryo's heart. Using time-lapse videos of more than 10,000 developing embryos, a deep learning system predicted the development of a heart[55] with an accuracy (AUC) of 0.98, compared to an accuracy of 0.74 by embryologists.

It is well known that antibiotics are becoming increasingly ineffective. Even though the application of neural networks to discovering novel antibiotics is not new, a recent breakthrough was made by using a neural network to predict the efficacy of drugs based on their molecular structure[46]. This resulted in a drug, *halicin*, which is effective against bacteria resistant to conventional antibiotics.

2.7 Language

Even though it is now relatively easy to get a neural network to learn to recognise pictures, it is much harder for such a system to provide a written account of what a picture shows (called *caption generation*), as in Figure 2.8.

Moreover, if we try to get an AI system to provide an extended written account of a series of pictures, or of a movie, then the underlying learning algorithms struggle or fail. The problem is that neural networks can work effectively only over relatively short spans of space (i.e. in a single image) or time (i.e. over a few letters or words). Neural networks are inherently *local* in time and space. This does not mean that they cannot be modified to see beyond the merely local, but such extensions are not a natural part of their intrinsic structure.

Additionally, the ability of neural networks to process language is hard to evaluate because, unlike vision, linguistic skill in translation (for example) is somewhat subjective. Despite such considerations, anecdotal tests of machine translation suggest that reasonable progress is being made.

A simple test of linguistic performance involves asking a neural network application such as *Google Translate* to translate an English text into another language, and then asking for the result to be translated back into English. For example, the text

Figure 2.8: For a picture like this, an AI system[3] generated the caption 'A man on a motorcycle going down a track'. Image reproduced with permission from Wikimedia under CCA License.

> *This book manages the impossible: it is a fun read, intuitive*
> *and engaging, lighthearted and delightful, and cuts right*
> *through the hype and turgid terminology.*

was translated to Arabic and then translated back into English using
Google Translate, which yielded

> *This impossible book manages the impossible: it is a*
> *pleasant, intuitive and attractive reading, beacon and*
> *cheerful, and it matches directly with the hype and trivial*
> *terms.*

This is not too impressive, but such applications seem to perform
better with text that is less abstract than this example. If this were
not true then the smart speaker assistant systems currently available
would not be able to make sense of human instructions or requests.

2.8 Is AI Biased?

It has been reported that AI systems perform poorly because they are
biased, especially with regard to non-white skin colour[31]. Such bias
is particularly relevant to applications involving skin colour, such as
face recognition (Section 2.2) and skin cancer diagnosis (Section 2.6).
There is no reason, in principle, to suppose that an AI system should
perform any better or worse on white skin. It therefore seems likely
that any bias is a result of training AI systems on data sets in which the
majority of images contain white skin, which would almost inevitably
lead to an AI system being more proficient at analysing images of
white skin. However, it should be noted that bias can be introduced
at several key stages, such as data collection, data processing, and
human categorisation of the data[32].

2.9 What AI Cannot Do

It used to be thought that games such as chess represented the pinnacle
of human cognitive ability, which may explain why symbolic AI was the
dominant form of AI from the 1950s until relatively recently. However,
the success of conventional symbolic AI systems reinforced a growing
suspicion that such systems find games like chess relatively easy. For
example, Deep Blue was a conventional computer program for chess
that won some games against the world chess champion Gary Kasparov

in 1997. Deep Blue performed a brute-force search for each move but did not employ any form of learning.

In contrast, the many failed attempts to get a robot to learn to perform tasks that a human child learns with ease suggests a dismal complementarity between humans and AI. For example, unlike humans, computers can do long division (and even play chess) with ease, whereas, unlike computers, humans can walk (and even run) up stairs with ease. In essence, tasks that are easy for humans are hard for computers, and vice versa, which is known as *Moravec's paradox*.

Given the natural propensity of neural networks to act coherently only over short spans of time and space, we should be unsurprised at the types of tasks at which they fail. A humanoid robot could be equipped with a variety of neural networks, each of which is specialised for individual domains such as vision, balance, hand control, etc. But despite some progress, the ability to learn tasks that require the integration of skills from several domains has yet to be achieved. For example, tasks on which no neural network has yet succeeded include:

- tying shoelaces,
- building a wardrobe from a flatpack,
- washing dishes,
- walking on rough ground,
- changing a tyre on a car,
- planting a sapling, and
- making a cup of tea.

Some of these examples might seem frivolous, but each represents a serious challenge to modern AI systems. Indeed, given that a typical adult can do *not just one but all* of these tasks, the challenge for modern AI systems looks formidable.

Chapter 3

How AI Works

3.1 Introduction

Almost all modern AI systems are essentially neural networks. There are many types of neural network, and correspondingly many types of learning algorithm, but the particular neural network that powers modern AI systems is the *backpropagation network*, usually abbreviated to *backprop* network. As its name suggests, a backprop neural network learns using the *backpropagation algorithm*, which is often combined with the *reinforcement learning algorithm*. The precise mix of backprop and reinforcement learning depends on the nature of the task being learned. For simple pattern recognition tasks such as classifying images (Figure 1.2), backprop can suffice. But for tasks involving sequences of actions, such as playing a game or trading shares, reinforcement learning is also required.

Learning in neural networks consists of a spectrum of methods, with *supervised learning* at one extreme and *unsupervised learning* at the other. Between these two extremes are two other types of method, *semi-supervised learning* and *self-supervised learning*. All of these make use of the backprop algorithm, and detailed accounts of all four types of learning are given below.

3.2 Supervised Learning

In supervised learning, a network is given both an input and a desired *target output*. The input is typically an image, and the output can be an image (as in Figure 1.3) or a *label* (as in Figure 1.2). For example, if the input is a picture of a cat, the output could be the word label *cat*, as in Figure 1.2. Because the correct output is given at the same time as the input, the network can work out how to change the connection weights between units to ensure that in future the response to this image will be the word *cat*. This learning is implemented using the backprop algorithm described in Section 3.4.

Supervised learning usually requires thousands of instances of *labelled data* for each class. For example, learning to distinguish between cats and dogs requires thousands of images that have each been labelled by hand as either *cat* or *dog*. A major obstacle for supervised learning systems is the limited availability of large amounts of labelled data. A common criticism of supervised learning is that neural networks require vastly more data than would be needed to teach a young child the difference between cats and dogs (for example).

Object Recognition. To understand how supervised learning works in a neural network, consider a relatively simple example. The task is to learn to discriminate between two classes of images, such as cats and dogs in Figure 1.2. More formally, the task is to learn to map each input to one of two output classes. Each input is an image that is a small 10-by-10 array of pixels. For the network to 'see' each image, a 10-by-10 array of input units is required, where each unit could correspond to a single photoreceptor in the eye. In order for the network to signal which class each image belongs to, there are two output units. For brevity, we label them C (for cat) and D (for dog). When an image contains a cat the network should switch unit C to its *on* state, and when an image contains a dog the network should switch unit D to its *on* state.

Between the *input layer* of 100 input units and the *output layer* of two output units, there are a fixed number of intermediate *hidden layers*, each of which can have a different number of units. Usually, all units in every layer are connected to the units in adjacent layers via connection *weights*. These weights correspond to the synapses between neurons in the brain, which are thought to underpin learning in physiological systems. It is these weights that define how each input image gets mapped to the output layer, and it is the job of the backprop algorithm to adjust each weight so that each input activates the correct output unit. And there's the rub.

Suppose the input image contains a cat but the D output unit state is *on* (indicating that the network thinks the image depicts a dog). This *classification error* occurs because one or more weights have the wrong value. However, it is far from obvious which weights are wrong and how they should be adjusted to correct the error. This general problem is called the *credit assignment problem*, but it should really be called the blame assignment problem, or even the whodunnit problem. In practice, determining whodunnit, and what to do with whoever-did-do-it, represents the crux of the problem to be solved by neural network learning algorithms. In order to learn, neural networks must solve this credit/blame assignment problem.

Just as the brain must (somehow) adjust synapses between neurons in order for us to learn, so backprop must adjust the weights between units for neural networks to learn. Unfortunately, and this is a very big unfortunately, adjusting the weights between units correctly is extremely hard — so hard that the problem of adjusting weights between units falls into a category called *NP-complete problems*. As far as we know, NP-complete problems are intractable.

This does not mean that sub-optimal solutions cannot be found using informed guesswork or reasonably powerful algorithms such as backprop. But knowing that a problem is NP-complete is a surefire signal that finding a solution will take time, and that any solution which is found is likely to be imperfect. Fortunately (and this is a big fortunately), in practice, if a massive number of training examples are available then neural networks perform well. Given the difficulty of the task faced by most neural networks, no-one is entirely sure why they work as well as they do. Indeed, much of the current research effort into neural networks is dedicated to elucidating this conundrum[40;51].

Innate Expectations. Whether a task involves pattern recognition or sequential actions, AI systems require vast amounts of data in order to gain sufficient experience to learn a particular task. However, the term vast, as used in the context of AI, does not really capture the sheer scale of the amount of data required. For example, the latest object recognition competition consisted of 14 million images from 2,000 classes. For comparison, a child needs exposure to only a few cats and dogs to be able to tell them apart.

The reason AI systems require so much data is probably that when an AI system is created, it has no particular expectations about the nature of the universe into which it has just been 'born'. This means that it has to learn everything from scratch, where the phrase 'from scratch' is accurately captured by the idea of a *tabula rasa*.

The term *tabula rasa* is often used in relation to human babies, but humans are born with innate expectations, or *priors*, about the fundamental nature of the physical world[48;59]. As a simple example, human babies expect the image on the retina to vary smoothly between adjacent photoreceptors, because corresponding adjacent points in the physical world usually have similar physical properties (e.g. colour, orientation, texture, depth). Such expectations are so basic, and so simple, that they are often overlooked.

But if babies did not expect the world to vary smoothly (for example), there would be no point in having a visual system that maps adjacent points in the world to adjacent points on the retina; in other words, there would be no point in forming a coherent image on the retina. Thus, the very existence of eyes that form coherent retinal images represents a form of expectation, or *Bayesian prior*[48], about the structure of the physical world. Even such a vague smoothness assumption provides an enormous head start over an AI system, which has to *learn* which pairs of units in its input layer should correspond to adjacent points in a coherent (e.g. retinal) image.

It is easy to dismiss the idea of building an AI system without a retina-like input layer as foolhardy. However, many of the earliest neural networks had to learn using arrays of pixels that were simply 'stretched out' versions of a two-dimensional camera image. In other words, to get a neural network to process a two-dimensional image of 10×10 pixels, consecutive 10-pixel rows were concatenated to produce a single row of 100 pixels, and this 100-pixel row was used as input to the neural network. Such mangling of visual input data places any neural network at an enormous disadvantage compared to a network that somehow knows that the underlying data are derived from a 10×10 image. In order to address this problem, LeCun introduced *convolutional neural networks* (see Section 3.3).

However, while emulating the basic architecture of the human visual system in neural networks may be necessary, it is clearly not sufficient. Even though convolutional neural networks require less data than conventional neural networks, they still require vast amounts of data. As if this were not enough bad news, there seems to be a general case of *diminishing returns* for each extra megabyte of training data[51]. In other words, every time the amount of training data is doubled, the amount of extra performance obtained shrinks; so more data is better, but not much better.

Fooling Neural Networks. Neural networks are fragile, in the sense that a well-chosen small change to an input can change the network's

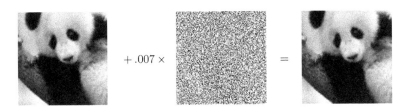

Figure 3.1: Fooling a neural network. The left image is classified as a panda with 58% confidence. If this image is corrupted by adding a specific noise image (centre), then the result (right) is classified as a gibbon with 99% confidence. With permission from Goodfellow et al. (2014).

output dramatically[13], as in Figure 3.1. Remarkably, the change to the input image is so small that it is invisible to human eyes.

3.3 Convolutional Neural Networks

Before the advent of convolutional neural networks[25] in the 1980s, most backprop neural networks simply ignored the two-dimensional structure implicit in image data. In contrast, convolutional neural networks were explicitly designed to mimic the structure of biological vision systems. Convolutional neural networks almost certainly account for the step-change in performance, particularly in terms of recognising handwritten characters.

In a conventional backprop neural network, every unit in every layer is connected to every unit in adjacent layers, as shown in Figure 1.2. In contrast, within a convolutional neural network, each unit in the first hidden layer collates information from a small region of the input layer, as shown in Figure 3.2. Consequently, if the network input is an image, then each hidden unit 'sees' only a small patch of the image. This mimics the human vision system, which has specialised neurons (*retinal ganglion cells*), each of which is connected to *photoreceptors* in a particular small region of the retina, called its *receptive field*.

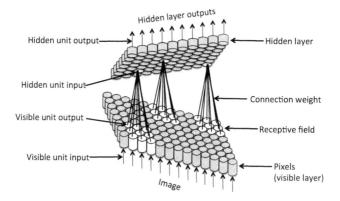

Figure 3.2: The first and second layers of a convolutional neural network. Each hidden unit receives connections from one region of the input layer.

Using hidden units that collate information from individual receptive fields results in a division of labour, such that each hidden unit concentrates on visual data within its own receptive field. This division of labour is repeated through successive layers, so that each hidden unit in the second hidden layer receives input from a small region of the first hidden layer, and so on.

After learning, it is often found that units in the early layers seem to be tuned to detect low-level features, such as contrast edges, whereas units in layers closer to the output layer seem to be tuned to detect high-order features, such as parts of objects. It is probably no coincidence that neurons in different parts of the human visual system are also tuned to increasingly abstract visual features.

3.4 The Backprop Algorithm

The backprop algorithm is the engine that powers modern AI systems. It is employed in all types of learning, from supervised learning to reinforcement learning. For the sake of tangibility, backprop is described for a basic supervised learning task here (see Figure 1.2).

Consider a training set consisting of a number of input/output pairs to be learned. For example, the input part of a pair could be an image of a cat or dog, and the output part could be a word (e.g. cat) that represents the desired or target output of the network (Figure 1.2). Each pair contributes a small amount to the learning process. For each pair, the backprop algorithm comprises two phases:

1. a *forward propagation phase*, in which the states of units in the first (input) layer propagate through successive hidden layers to the final (output) layer; and

2. a *backward propagation phase*, in which errors at the final layer propagate backwards to the first layer.

During the backward propagation phase, the amount of error is calculated for each unit, for each input/output pair. This error is then used to calculate a small change for each of that unit's connection weights, where this change improves performance of the neural network on that input/output pair. A learning iteration, or *epoch*, consists of presenting all of the input/output pairs in the training set and calculating the changes to each weight. Learning, which typically requires many thousands of epochs, is terminated when the network performance is deemed satisfactory (see Section 3.9).

3.5 Semi-Supervised Learning

As mentioned above, a major problem for supervised learning systems is the limited availability of large amounts of labelled data. Two important types of neural networks that circumvent, or at least alleviate, this problem are the *autoencoder* and the *generative adversarial network* (GAN).

The autoencoder was originally introduced in 1984 as a *Boltzmann machine* neural network[15]. Modern incarnations are the *variational autoencoder*[9;22] (VAE) and the *deterministic autoencoder*[12]. Autoencoders use each image in a data set as both the input and the target output, as shown in Figure 3.3. This might sound a bit odd, but the idea is to force the network to transmit information from the input layer to the output layer via a small number of units in a middle layer. This middle layer acts like a bottleneck, forced to extract as much information as possible about the data. Because the output of the middle layer provides a compact representation of the image presented to the input layer, it can then be used as the input to a supervised neural network. This compactness usually means that each unit in the middle layer represents a high-level feature, which

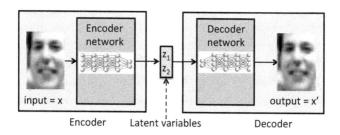

Figure 3.3: Semi-supervised learning using a variational autoencoder, which comprises an encoder and a decoder. The encoder maps each input image (x) to several output unit states (z_1 and z_2). These act as inputs to the decoder, which produces an approximation (x') of the input image.

ensures that the supervised neural network requires relatively small amounts of labelled data[1;23;35].

A generative adversarial network[13] (GAN) consists of two separate backprop networks, a *generator network* and a *discriminator network*, as shown in Figure 3.4. The discriminator is presented with images from the generator or from a training set of images taken by a camera. On each trial, the generator assigns a probability to its input image, according to how realistic the image appears to be. In a sense, these two networks are engaged in an arms race, in which the generator gradually learns to produce realistic synthetic images, while the discriminator simultaneously learns to differentiate between the generator's synthetic images and images from the training set.

Such an adversarial framework has also been extended to employ autoencoders[34], and it can be used to generate high-resolution images of synthetic celebrities, as shown in Figure 2.4.

3.6 Self-Supervised Learning

Self-supervised learning is essentially a method for forcing a neural network to learn the underlying grammar or statistical structure of

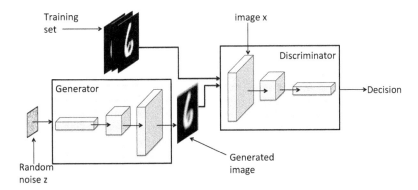

Figure 3.4: A generative adversarial network (GAN) comprises two networks, a generator and a discriminator. The discriminator receives an image and has to learn whether that image is from the generator or from a training set of images taken by a camera. The generator learns to generate increasingly realistic images to fool the discriminator.

a data set, based on the idea of natural pairings between input and output items. The initial motivation for using self-supervised learning was that there is a finite amount of labelled data in the world.

To increase the amount of data available for training, a network can first be trained on a *pretext task*. A common example of a pretext task is to use one or more images from a movie as the network input, and an image from several seconds later as the target output. In other words, the network's task is to predict the image that will occur in the near future. In order for the network to successfully learn to predict images several seconds in the future, it must learn the underlying spatiotemporal statistical structure of the movie data set.

After learning on a pretext task, training on a small amount of labelled data achieves levels of performance that would otherwise require large numbers of input/output training pairs that have been labelled by hand. The idea of self-supervised learning has been around at least since 1990[37], but there has recently been a resurgence of interest in this method[18;28], which has exhibited the best object recognition performance to date[8].

3.7 Unsupervised Learning

The supervised learning described above involves an input together with a desired output, which is shown to the network and effectively acts as a teacher. In contrast, with unsupervised learning, a network learns to sort raw data into different categories without a teacher or any other form of feedback regarding the accuracy of its outputs. For example, if the input is a mixture of two sets of pictures, one set containing cats and the other containing dogs, then the network would be expected to sort the pictures such that its output provides a distinction between the set of cat pictures and the set of dog pictures.

Unsupervised learning relies on the existence of some intrinsic similarity between items in the same class. It is often said that unsupervised learning 'lets the data speak for itself'. However, all unsupervised learning methods depend on biases implicit in their design, even if these biases are not obvious to the methods' designers.

To take an extreme example, if the data consist of pictures of different animal species, and if the learning algorithm has an implicit bias to classify animals according to the number of legs they have, then it is entirely reasonable for humans to be grouped with penguins. On the other hand, if animals were grouped according to intelligence, it would be reasonable to group humans with whales. More typically, unsupervised learning depends on low-level features of inputs, so such algorithms tend to group inputs according to low-level properties like image brightness. Of course, if the inputs to a neural network are not images but high-order features such as disease symptoms, then (ideally) similar symptoms would be grouped into classes that correspond to similar diseases.

Before AI became synonymous with neural networks, it dealt almost exclusively with *symbolic representations*, such as the word *cat*, and conditional relations, such as *if a cat sees a mouse then* .... It soon became apparent to researchers in symbolic AI that once the data have been recoded appropriately, then to all intents and purposes the underlying problem has been solved. In other words, the hard part of solving most problems consists of identifying key properties or *features* that allow superficially dissimilar items to be recognised as similar. Likewise, for a neural network, if data are recoded by a human to make subtle properties explicit and these properties are then presented as inputs to the neural network, the problem becomes trivial. For example, instead of presenting a neural network with raw images, certain visual features (e.g. the number of legs in each image) could be identified by a human operator, and each feature could then be presented to the neural network as a single variable. This is not meant to imply that neural networks can solve hard problems only if those problems have been rendered trivial by judicious recoding of their inputs into high-level features, but rather to emphasize that the hard part of solving most problems is discovering those high-level features from raw data.

Traditional methods for classifying data without the aid of an explicit teacher include *principal component analysis* (PCA), *factor analysis* (FA) and *independent component analysis*[47] (ICA). These methods are limited because they rely on low-level features (e.g. groups of pixels), but it is high-order features that are required to classify objects correctly.

3.8 Reinforcement Learning

Animals choose actions that are rewarding and avoid actions that have negative consequences. The feedback received from actions can be used to learn which actions are most rewarding, and the process of learning to maximise rewards on the basis of feedback is called *reinforcement learning*. This general idea has been refined over many years into a suite of *reinforcement learning algorithms*[52], which use trial and error to learn a sequence of actions that maximises total reward. Reinforcement learning is an inspired fusion of game playing by computers, as developed by Shannon (1950) and Samuel (1959), optimal control theory, and stimulus–response experiments in psychology. Skills learned using reinforcement learning include playing chess, backgammon, Go (see Section 2.4) and pong (Figure 2.5), balancing a pole (Figure 2.6), and learning to soar (Figure 2.7).

The problem confronting any learning system (biological or artificial) is that the consequences of each action often occur some time after the action has been executed. Meanwhile, many other actions could have been executed, so disentangling how much each action contributed to a later reward is a difficult problem. For example, if a rat eats tainted food then nausea usually follows several hours later, during which time the rat has probably eaten other items, so it is hard to be certain which item caused the nausea (actually, rats are very good at identifying which food item caused nausea). Similarly, the final outcome of a chess game depends on every move within the game, so the wisdom of each move can be evaluated only after the game is over. Evaluating the cost or benefit of each action lies at the heart of optimising behaviour, and is called the *temporal credit assignment problem*.

The ability to retrospectively assign credit to each action would be useless if a neural network could not use the benefits of hindsight to foresee the likely outcome of its current actions. It follows that a neural network's ability to solve the temporal credit assignment problem involves predicting future states of the world around itself. Fortunately, it turns out to be much easier to predict the future if that future is determined by, or at least affected by, the neural network's own actions. By implication, the ability of a neural network to take actions that maximise rewards depends heavily on its ability to predict the outcome of its actions.

Historically, the origins of reinforcement learning algorithms can be traced back to Shannon (1950), who described a program for playing chess. Even though his program was not designed to learn, he proposed an idea that to all intents and purposes is reinforcement learning:

> *a higher level program which changes the terms and coefficients involved in the evaluation function depending on the results of games the machine has played.*

Indeed, this idea seems to have been the inspiration for Samuel (1959), who designed a program for learning to play draughts (checkers). A later version of Samuel's program beat the checkers champion of the U.S. state of Connecticut in 1961. Since that time, reinforcement learning has been developed by many scientists, but its modern incarnation is due principally to Sutton and Barto (2018).

3.9 Over-Fitting

After training a neural network on a given data set, the network as a whole represents a mapping (a mathematical function) from the network's input to its output. Ideally, the training process results in a

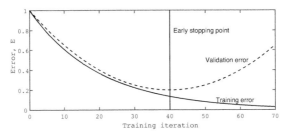

Figure 3.5: Reducing over-fitting by early stopping (see text).

network function that is a good estimate of the underlying trend in the data. However, if the training set is too small, the network function may fit every data point almost exactly, rather than fitting the general trend in the data. Such *over-fitting* is undesirable because it means that any predictions depend not only on the underlying trend in the data but also on the particular idiosyncrasies of the training set used. For neural networks, over-fitting can be reduced by *regularisation*.

A simple and robust regularisation method is *early stopping*. This consists of splitting the data into three sets: a *training set* (90% of the data), a *validation set* (5%), and a *test set* (5%). The network is trained on the training set, while performance is monitored on the validation set (Figure 3.5). Once the performance error on the validation set begins to increase, this indicates that over-fitting of the training set is occurring, so training is stopped. The test set is not used during training. Its only purpose is to estimate how well the network would perform when faced with new data; in other words, it tests how well the network generalises beyond the training data.

3.10 Is AI Just Curve Fitting?

A common criticism of AI with regard to supervised learning is that it is just a form of curve fitting, formally known as *regression analysis*.

Regression analysis works like this. Suppose we want to predict the income of a person, John, when he reaches the age of 40. To do this, we begin by collating information from thousands of individuals who are currently over 40. For each individual, we record their income at different ages. Regression analysis consists, in essence, of fitting a curve to these data points, as shown in Figure 3.6. We can find John's predicted income at age 40 by drawing a line up from the x-axis (Age) at the point marked 40 until this line hits the curve, and then drawing a horizontal line to the y-axis (Income) and reading off the value there; in this example, John's predicted income at age 40 is $74,000.

Of course, income depends on more than just age, so we should really take account of other factors, such as IQ, college grades, and so on. Accordingly, if we measure these variables for a large number of people then we can fit a curve to their data. In general, this curve will be a multi-dimensional surface, so we cannot draw it; but it can be used in the same way as above. So, given John's college grades, IQ etc., we can use this curve as a kind of continuous look-up table to read off his predicted income at age 40. We have omitted many technical details here, but that, basically, is regression analysis.

Similarly, if we have data from many thousands of individuals, we could use the age, IQ, college grades etc. of each individual as a set of inputs to a backprop neural network, and use the income of each individual as the network's target output. Then, when John's characteristics (age, IQ etc.) are given as input to the network, the output will represent the network's prediction of his income at 40. The problem is the same whether it is expressed in terms of regression analysis or in terms of AI (specifically, a neural network). But AI and regression analysis differ in three important respects.

First, AI seems to have the ability to fit (what appears to be) the correct curve for millions of data points; in other words, AI systems consistently outperform regression analysis. Formally, this curve represents a complicated mathematical function that maps neural network inputs to outputs. During learning, neural networks adjust the weights between units so that the overall mathematical function represented by the network provides accurate predictions.

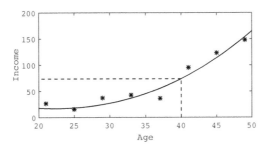

Figure 3.6: Scelmatic example of curve fitting for age and income (\$'000). Each asterisk represents the mean salary of thousands of people at a given age. The black line is a curve that has been fitted to the data points, and which represents the underlying trend of how salary increases with age.

Second, whereas regression analysis requires hand-crafted quantities (e.g. IQ) to be used as input variables, AI systems can use raw data (e.g. images and sounds) as inputs. This relieves human programmers from the hard task of recoding raw data into high-level variables (features) which it is hoped would be relevant to the task at hand. More importantly, because AI systems derive their own features from raw data to solve a given task, they tend to extract features that are best suited to solving that task. In other words, the features derived and used by AI systems are usually superior to hand-crafted features chosen by a human operator.

Third, many tasks (e.g. playing chess) cannot be formulated as regression problems. These tasks require some form of reinforcement learning (Section 3.8) and cannot be solved using any type of 'curve fitting', either with conventional regression analysis or by using a backprop neural network on its own.

In conclusion, to describe AI as *just* regression analysis is a bit like saying that a Harrier jump jet is just a type of Wright brothers plane, which may be true but is entirely unhelpful for understanding how a modern jet plane functions. It is also a bit like saying that a computer is just a type of abacus, which is true but no help at all in understanding how a computer works. In both cases, the word *just* is doing an unreasonable amount of work. A jet plane is a type of Wright brothers plane, but it is not *just* a Wright brothers plane; a computer is a type of abacus, but it is not *just* an abacus.

The fact that AI can outperform conventional regression analysis without the need for hand-crafted inputs (features) suggests that AI is not *just* regression analysis. More generally, many AI systems, particularly those that include various combinations of backprop and reinforcement learning, do not *just* perform regression analysis, because what they do is nothing like regression analysis.

Chapter 4

The Future of AI

4.1 Introduction

The future of AI has been compared to the future of nuclear fusion. The claim that nuclear fusion would provide free, unlimited power within ten years was first made soon after it was discovered in 1930, and that claim has been renewed about every ten years since then. Similarly, a promise that AI would exist within ten years was first made in the initial wave of AI research in the 1950s, and that promise has been renewed on a fairly regular basis ever since. For example, Claude Shannon, the inventor of *information theory*[42;49], said in 1961:

> *I confidently expect that, within 10 to 15 years, we will find emerging from the laboratory something not too far from the robots of science fiction fame.*

If (as is argued below) AI will one day become a reality, then continually renewing a prediction that AI will succeed within ten years is a safe bet, because at some point that prediction must come true.

However, before we consider the question of whether computers will ever be intelligent, we should address the following fundamental question: is AI possible, even in principle? The answer is *yes*, and here's why.

Before Orville and Wilbur Wright flew the first aeroplane in 1903, sceptics declared that a machine could never fly like a bird*. Today,

---

*In 1901, Wilbur had predicted that powered flight would take another 50 years.

many of us are like those sceptics, doubting that a machine can ever achieve human levels of intelligence. But for a compelling counter-argument to such scepticism, consider birds and brains as physical devices that must obey the same universal laws of physics. In other words, a bird is a flying machine that happens to be made of organic matter, and a brain is a computational machine that happens to be made of neurons.

From this perspective it seems obvious, and even inevitable, that a machine should be able to fly even if it is not made of organic matter, and that a machine can be intelligent even if it is not made of neurons. If we accept that the brain is a type of machine, then:

> The question of whether machines can think is about as relevant as the question of whether submarines can swim.
> Edsger Dijkstra, 1984.

Indeed, the most succinct and conclusive answer to the question of whether a machine can think is to be found by looking in a mirror.

4.2 Time Enough For AI?

Regarding the question of when artificial intelligence will arrive, consider this. Whereas it took millions of years for the first gliding vertebrate to evolve into a bird that could fly, it took a mere 66 years from the Wright brothers' first flight to Neil Armstrong's 'small step' onto the moon in 1969. Similarly, even though life probably appeared on Earth as soon as it cooled, it took a further four billion years for the first animals with primitive neurons to evolve, but a mere half a billion years more for natural selection to sculpt those first neurons into the human brain.

As a final example, the first electronic computers were made of vacuum tubes (like old-style tungsten bulbs). To give some idea of the rate of progress, and of our failure to appreciate how quickly progress accelerates, a 1949 article in *Popular Mechanics* magazine declared:

> *... a calculator today is equipped with 18,000 vacuum tubes*
> *and weighs 30 tons, computers in the future may have only*
> *1,000 vacuum tubes and perhaps weigh only half a ton.*

The first gliding vertebrates could not imagine eagles; the Wright brothers could not imagine humans walking on the moon within 70 years of their historic flight; the 1949 writer for *Popular Mechanics* could not imagine an entire computer inside a phone. Similarly, we almost certainly cannot imagine how progress will have changed this world 50 or 100 years into the future. To do so would be remarkable because, as we have seen from the examples above, the rate of progress seems to accelerate, making long-term prediction almost impossible. In conclusion, the examples above suggest that, within both natural and human-influenced domains, progress starts slowly but then accelerates at a rate that may be exponential. And if progress really is accelerating exponentially, then the answer to the question of when artificial intelligence will arrive is — *soon.*

4.3 Another Revolution?

In its early stages, every technological revolution gets compared to the emperor's new clothes. For example, neural networks are often described as just a type of dressed-up regression analysis (see Section 3.10). In an otherwise cogent critique of deep learning, Marcus (2018) describes deep learning as 'just mathematics' and 'just a statistical technique'. Other common criticisms are that the achievements of modern neural networks are due to *just faster computers*, or *just more data*, or *just both* — as if no other ingredient is required (such as sustained research effort). Such criticisms come from the nay-sayers of the revolution.

Then there are the cheerleaders, who react as if time-travel has just been invented. Books with exaggerated titles are written, debates with simplified binary choices are held in front of eager audiences, battle lines are drawn, heroes are identified, and long-forgotten soothsaying geniuses are rediscovered. However, this frenzy of hyperbole effectively hides whether or not there is a revolution and, if there is, what the revolutionaries have discovered that is so revolutionary.

For example, a common theme of the binary debates is whether or not artificial intelligence will destroy the human species. One reason for the popularity of such debates is that having an opinion about artificial intelligence is easy; it does not involve having to know anything much, and (best of all) it does not require having to learn anything new or hard. For the record, I think it is improbable that artificial intelligence will destroy the human species, but the honest answer is that no-one knows for sure. Despite this, almost anyone with or without a PhD in nothing-to-do-with artificial intelligence seems to have a firm opinion on such matters.

4.4 Machine Consciousness

The prospect of AI inevitably raises the question of consciousness. But the problem with the question of consciousness is that there is no consensus about exactly what the question is. And, whatever it is, there is no objective measure that can be used as a test for the existence, or the amount, of consciousness. I am fairly sure other humans are conscious, and I am absolutely certain that my cat is conscious, but I have no objective evidence in either case.

Such an interesting and ineffable question naturally generates much discussion, but little in the way of clarity. To some extent, the question of whether or not a neural network can become conscious can be answered by the following exercise. Take one (presumably conscious) human brain with its 86 billion neurons. Choose one neuron at random, and replace it with a microprocessor (an artificial neuron) that has been designed to have exactly the same input/output characteristics as the chosen neuron. Now repeat that 86 billion times.

The result is a silicon neural network, a brain that has exactly the same functionality as the biological brain it replaced. Is this silicon brain conscious? The question is as unanswerable for the silicon brain as it is for the biological brain. In both cases, we have no evidence. However, the lack of evidence for the biological brain (which we assume is conscious) is the same lack of evidence as for the silicon brain. This suggests that there is nothing special about biological neural networks that allows them to give rise to consciousness.

In any case, the debate regarding whether or not neural networks are conscious is, at the very least, premature. Aside from some superficial similarities, artificial neural networks have very little in common with biological neural networks, and the experiment described above cannot anyway be attempted with current technology.

Geoffrey Hinton is one of the pioneers of modern AI, so his thoughts on such matters should not be ignored. Around the year 2002, I attended one of his lectures. At the end, a member of the audience asked Hinton whether or not neural networks would ever become conscious[†]. Hinton replied with this parable[‡] (paraphrased from memory):

> *In England, when two people get married, tin cans are tied to the newly weds' car with long pieces of string, so that the cans make plenty of noise as they get dragged along the road. It is the car that provides the motive force but it is the cans that make all the noise.*

Hinton is English, so this was his polite way of explaining that *there are scientists who do research, and there are academics who discuss what the scientists who do research, do.*

Note that it has not been claimed that neural networks cannot be conscious, simply that it is too early to spend time investigating this topic. Indeed, given that the human brain is essentially a biological machine, the answer to the question of whether a machine can be conscious is the same as the answer to the question of whether a machine can think, and neither question should require much time to answer. But the day will come (and may have already arrived) when an elderly scientist declares that a conscious AI is impossible.

[†]To be honest, I cannot recall if the question was about intelligence or consciousness, but the answer would have been the same in both cases.

[‡]This is a story originally told by the physicist John Wheeler. During a lecture on General Relativity, he drew a car with cans trailing behind, and said, "The car is like physics and the cans are like the philosophers: When the car moves forward the cans make all the noise". (T Sejnowski, personal communication).

On that day, it will be worth reminding ourselves of the words of an eminent science fiction writer:

> *When a distinguished but elderly scientist states that something is possible, he is almost certainly right. When he states that something is impossible, he is very probably wrong.*

Arthur C Clarke's First Law, 1962.

Further Reading

Most of the items listed here are freely available online.

Bishop, CM (2006). *Pattern Recognition and Machine Learning.* Oxford University Press. Technical and comprehensive account of neural networks up to 2006, and has a strong Bayesian perspective. https://www.microsoft.com/en-us/research/people/cmbishop/prml-book/

Goodfellow, B, Bengio, Y, and Courville, A (2016). *Deep Learning.* MIT Press. A comprehensive reference text on neural networks.

Hinton, G and LeCun, Y (2019). *Turing Lecture*, June 23, 2019, Federated Computing Research Conference.
An overview from the founders of modern AI.
https://www.youtube.com/watch?v=VsnQf7exv5I

Marcus, G (2018). Deep Learning: A Critical Appraisal.
A comprehensive and cogent critique of modern AI by an author with substantial hands-on experience.
https://arxiv.org/abs/1801.00631

Nielsen, M (2015), *Neural Networks and Deep Learning.*
A little dated, but still provides a fine starting point for learning about the backpropagation learning algorithm.
http://neuralnetworksanddeeplearning.com

Sejnowski, TJ (2018). *The Deep Learning Revolution.* MIT Press.
A fine mixture of historical overview and memoir from one of the pioneers of neural networks.

Silver, D (2018). *Reinforcement Learning* lecture slides.
Succinct but, more importantly, reliable and comprehensive account of reinforcement learning algorithms; highly recommended.
http://www0.cs.ucl.ac.uk/staff/d.silver/web/Teaching.html

Further Reading

Stone, JV (2019). *Artificial Intelligence Engines: A Tutorial Introduction to the Mathematics of Deep Learning.* Sebtel Press. https://jim-stone.staff.shef.ac.uk/AIEngines

Sutton, RS and Barto, AG (2018). *Reinforcement Learning: An Introduction.* MIT Press (available free online). Comprehensive account by two of the masters of modern reinforcement learning. http://incompleteideas.net/book/the-book-2nd.html

Zhang, A, Lipton, ZA, Li, M, and Smola, AJ (2020). *Dive into Deep Learning.* A comprehensive (852 pages) and well-written book. https://d2l.ai

Glossary

accuracy Given items to be classified as belonging to either class X or class Y, the accuracy A is the average of the true positive rate (TP), which is the proportion of items from X correctly classified as X, and the true negative rate (TN), which is the proportion of items from Y correctly classified as Y. So A = (TP + TN)/2.

algorithm A definite method or sequence of steps.

AUC A robust measure of classification performance. It stands for *area under the curve* of a receiver operating characteristic (ROC) curve. An AUC of 0.5 means classification performance is at chance level, and an AUC of 1.0 means classification is perfect. Under mild assumptions, AUC increases with *accuracy*.

autoencoder A neural network trained to map inputs to identical outputs via a 'bottleneck', a small number of intermediate units in a central hidden layer.

backprop Short for backpropagation, it is the principal algorithm for training neural networks.

content addressable memory If an association between an input and an output is learned, and if the input evokes that output, then the memory is said to be content addressable.

convolutional neural network A network in which each unit in the first hidden layer collates information from a small region of the input layer.

deepfake A synthetic version of a real person speaking, where the speech is fabricated by the creator of the synthetic version.

deep learning Learning in a deep neural network, which usually relies on a form of *backprop*.

deep neural network A neural network with more than one hidden layer of units between the input and output layers.

generalisation The ability to extrapolate beyond training data items.

generative adversarial network A system of two competing networks in which a generator network produces outputs, and a discriminator network decides whether each output is from the generator network or from the training set.

interpretability Ability to infer which aspects of AI inputs are most relevant to its responses or outputs.

labelled data Data where the correct output for each input is known.

machine learning An umbrella term that encompasses neural networks and reinforcement learning.

Moravec's paradox Tasks that computers find easy (e.g. long division) humans find hard, and tasks that humans find easy (e.g. tying laces) computers find hard.

over-fitting Given noisy data with an underlying trend, an ideal model would fit a function to the trend. Over-fitting occurs when the model fits a function to individual data points.

regularisation A class of methods designed to reduce over-fitting.

reinforcement learning A class of methods that use trial and error to learn a sequence of actions that maximises total reward.

self-supervised learning Learning based on natural pairings of input and output items. For example, all pairs of consecutive images in a movie are natural pairs, where the first member of each pair can act as the input to a neural network and the second member can act as the desired output.

semi-supervised learning Learning the underlying statistics of a data set by using the each data item as both the input and the output of a neural network.

supervised learning Learning the mapping from a set of inputs to a corresponding set of outputs.

symbolic AI AI that depends only on operations between high-level symbols (e.g. if see food then eat). Symbolic AI dominated AI until about 2010; also known as *good old fashioned AI* (GOFAI).

training data Data used to train a neural network, usually split into three sets (training, validation, and test sets).

unit Artificial neuron.

unsupervised learning Learning with unlabelled data. If the data contain well-defined clusters, an unsupervised learning algorithm should represent the data in terms of these clusters.

weight The strength of the connection between two units.

Bibliography

[1] AA Alemi, I Fischer, JV Dillon, and K Murphy. Deep variational information bottleneck. *arXiv e-prints*, 2016. arXiv:1612.00410.

[2] M AlQuraishi. A watershed moment for protein structure prediction, 2020.

[3] C Amritkar and Vi Jabade. Image caption generation using deep learning technique. In *Proceedings of the Fourth International Conference on Computing Communication Control and Automation (ICCUBEA)*, pages 1–4. IEEE, 2018.

[4] C Badue et al. Self-driving cars: A survey. *arXiv e-prints*, 2019. arXiv:1901.04407.

[5] GA Bello et al. Deep-learning cardiac motion analysis for human survival prediction. *Nature Machine Intell.*, 1(2):95–104, 2019.

[6] DC Berry and DE Broadbent. Interactive tasks and the implicit-explicit distinction. *British J. of Psychology*, 79(2):251–272, 1988.

[7] C Bishop. *Pattern Recognition and Machine Learning*. Springer, 2006.

[8] T Chen, S Kornblith, M Norouzi, and G Hinton. A simple framework for contrastive learning of visual representations. *arXiv e-prints*, 2020. arXiv:2002.05709.

[9] C Doersch. Tutorial on variational autoencoders. *arXiv e-prints*, 2016. arXiv:1606.05908.

[10] A Esteva, B Kuprel, RA Novoa, J Ko, SM Swetter, HM Blau, and S Thrun. Dermatologist-level classification of skin cancer with deep neural networks. *Nature*, 542(7639):115–118, 2017.

[11] BH Gary, R Manu, B Tamara, and LM Erik. Labeled faces in the wild: A database for studying face recognition in unconstrained environments. Technical report 07-49, University of Massachusetts, Amherst, October 2007.

[12] P Ghosh, MSM Sajjadi, A Vergari, M Black, and B Schölkopf. From variational to deterministic autoencoders. *arXiv e-prints*, 2019. arXiv:1903.12436.

[13] IJ Goodfellow, J Pouget-Abadie, M Mirza, B Xu, D Warde-Farley, S Ozair, A Courville, and Y Bengio. Generative adversarial networks. *arXiv e-prints*, 2014. arXiv:1406.2661.

[14] I Guilliard, R Rogahn, J Piavis, and A Kolobov. Autonomous thermalling as a partially observable Markov decision process. *arXiv e-prints*, 2018. arXiv:1805.09875.

[15] GE Hinton, TJ Sejnowski, and DH Ackley. Boltzmann machines: Constraint satisfaction networks that learn. Technical report, Dept. Computer Science, Carnegie-Mellon University, 1984.

[16] M Hutson. From models of galaxies to atoms, simple AI shortcuts speed up simulations by billions of times. *Science*, 367(6479):728, 2020.

[17] W Jiao et al. A deep learning system can accurately classify primary and metastatic cancers based on patterns of passenger mutations. *BioRxiv*, 2019. doi: 10.1101/214494.

[18] L Jing and Y Tian. Self-supervised visual feature learning with deep neural networks: A survey. *arXiv e-prints*, 2019. arXiv:1902.06162.

[19] T Karras, T Aila, S Laine, and J Lehtinen. Progressive growing of GANs for improved quality, stability, and variation. *arXiv e-prints*, 2017. arXiv:1710.10196.

[20] B Kim, KR Varshney, and A Weller. Proceedings of the 2018 ICML Workshop on Human Interpretability in Machine Learning (WHI 2018). *arXiv e-prints*, 2018. arXiv:1807.01308.

[21] E Kim, H Kim, K Han, BJ Kang, YM Sohn, OH Woo, and CW Lee. Applying data-driven imaging biomarker in mammography for breast cancer screening: Preliminary study. *Scientific Reports*, 8(1):1–8, 2018.

[22] DP Kingma, S Mohamed, DJ Rezende, and M Welling. Semi-supervised learning with deep generative models. In *NIPS'14: Proc. 27th International Conference on Neural Information Processing Systems*, pages 3581–3589. MIT Press, 2014.

[23] DP Kingma and M Welling. Auto-encoding variational Bayes. *arXiv e-prints*, 2013. arXiv:1312.6114.

[24] CP Langlotz. Will artificial intelligence replace radiologists? *Radiology: Artificial Intelligence*, 1(3), 2019. doi: 10.1148/ryai.2019190058.

[25] Y LeCun, B Boser, JS Denker, RE Henderson, W Hubbard, and LD Jackel. Backpropagation applied to handwritten ZIP code recognition. *Neural Computation*, 1(4):541–551, 1989.

[26] G Marcus. Deep learning: A critical appraisal. *arXiv e-prints*, 2018. arXiv:1801.00631.

[27] MultiMedia LLC. Obama deep fake, 2017.

[28] I Misra and L van der Maaten. Self-supervised learning of pretext-invariant representations. *arXiv e-prints*, 2019. arXiv:1912.01991.

[29] V Mnih, K Kavukcuoglu, D Silver, A Graves, I Antonoglou, D Wierstra, and M Riedmiller Playing Atari with deep reinforcement learning. *arXiv e-prints*, 2013. arXiv:1312.5602.

[30] V Mnih et al. Human-level control through deep reinforcement learning. *Nature*, 518(7540):529–533, 2015.

[31] P Noor. Can we trust AI not to further embed racial bias and prejudice? *BMJ*, 368, 2020. doi: 10.1136/bmj.m363.

[32] E Ntoutsi et al. Bias in data-driven AI systems – an introductory survey. *arXiv e-prints*, 2020. arXiv:2001.09762.

[33] H Pham, MY Guan, B Zoph, QV Le, and J Dean. Efficient neural architecture search via parameter sharing. *arXiv e-prints*, 2018. arXiv:1802.03268.

[34] A Radford, L Metz, and S Chintala. Unsupervised representation learning with deep convolutional generative adversarial networks. *arXiv e-prints*, 2015. arXiv:1511.06434.

[35] DJ Rezende, S Mohamed, and D Wierstra. Stochastic backpropagation and approximate inference in deep generative models. *arXiv e-prints*, 2014. arXiv:1401.4082.

[36] AL Samuel. Some studies in machine learning using the game of checkers. *IBM J. Research and Development*, 3(3):210–229, 1959.

[37] J Schmidhuber. Making the world differentiable: On using self-supervised fully recurrent neural networks for dynamic reinforcement learning and planning in non-stationary environments. Technical report, Institut für Informatik, Technische Universität München, 1990.

[38] F Schroff, D Kalenichenko, and J Philbin. FaceNet: A unified embedding for face recognition and clustering. In *Proceedings of the 2015 IEEE Conference on Computer Vision and Pattern Recognition (CVPR)*, pages 815–823. IEEE, 2015.

[39] TJ Sejnowski. *The Deep Learning Revolution.* MIT Press, 2018.

[40] TJ Sejnowski. The unreasonable effectiveness of deep learning in artificial intelligence. *Proceedings of the National Academy of Sciences*, 2020. doi: 10.1073/pnas.1907373117.

[41] CE Shannon. Programming a computer for playing chess. *The London, Edinburgh, and Dublin Philosophical Magazine and Journal of Science*, 41(314):256–275, 1950.

[42] CE Shannon and W Weaver. *The Mathematical Theory of Communication.* University of Illinois Press, 1949.

[43] D Silver et al. Mastering the game of Go with deep neural networks and tree search. *Nature*, 529:484–503, 2016.

[44] D Silver et al. Mastering the game of Go without human knowledge. *Nature*, 550(7676):354–359, 2017.

[45] JD Smith et al. Implicit and explicit categorization: A tale of four species. *Neuroscience & Biobehavioral Reviews*, 36(10):2355–2369, 2012.

[46] JM Stokes et al. A deep learning approach to antibiotic discovery. *Cell*, 180(4):688–702.e13, 2020.

[47] JV Stone. *Independent Component Analysis: A Tutorial Introduction*. MIT Press, 2004.

[48] JV Stone. *Bayes' Rule: A Tutorial Introduction to Bayesian Analysis*. Sebtel Press, 2013.

[49] JV Stone. *Information Theory: A Tutorial Introduction*. Sebtel Press, 2015.

[50] JV Stone. *Artificial Intelligence Engines: A Tutorial Introduction to the Mathematics of Deep Learning*. Sebtel Press, 2019.

[51] C Sun, A Shrivastava, S Singh, and A Gupta. Revisiting unreasonable effectiveness of data in deep learning era. In *Proceedings of the 2017 IEEE International Conference on Computer Vision (ICCV)*, pages 843–852. IEEE, 2017.

[52] RS Sutton and AG Barto. *Reinforcement Learning: An Introduction*. MIT Press, 2018.

[53] S Suwajanakorn, SM Seitz, and I Kemelmacher Shlizerman. Synthesizing Obama: Learning lip sync from audio. *ACM Transactions on Graphics*, 36(4):1–13, 2017.

[54] EJ Topol. High-performance medicine: The convergence of human and artificial intelligence. *Nature Med.*, 25(1):44–56, 2019.

[55] D Tran, S Cooke, PJ Illingworth, and DK Gardner. Deep learning as a predictive tool for fetal heart pregnancy following time-lapse incubation and blastocyst transfer. *Human Reproduction*, 34(6):1011–1018, 2019.

[56] X Wang, et al. Chestx-ray8: Hospital-scale chest x-ray database and benchmarks on weakly-supervised classification and localization of common thorax diseases. In *Proceedings of the IEEE Conference on Computer Vision and Pattern Recognition*, pages 2097–2106. IEEE, 2017.

[57] H Wu and X Gu. Towards dropout training for convolutional neural networks. *Neural Networks*, 71:1–10, 2015.

[58] Q Yan, DE Weeks, H Xin, A Swaroop, EY Chew, H Huang, Y Ding, and W Chen. Deep-learning-based prediction of late age-related macular degeneration progression. *Nature Machine Intelligence*, 2(2):141–150, 2020.

[59] AM Zador. A critique of pure learning and what artificial neural networks can learn from animal brains. *Nature Communications*, 10(1):1–7, 2019.

Index